THE POULTRYMAN'S FORMULARY

DR. PRINCE T. WOODS
MIDDLETON, MASS., U. S. A.

Table of Contents.

CHAPTER I.
BALANCED-RATION FOODS.

Whole Grain Hopper Mixtures; Exercise or Scratch-Grain Foods; Dry-Mash Mixtures; Moist Mashes and Laying Food; Chick Foods; Growing Food; Fattening Foods; Home-made Meat Food. Formulaes 1-26. Pages 8-12

CHAPTER II.
DUCK RATIONS.

Rations for Breeding Ducks and Layers; Growing Rations for Ducklings; Finishing Off Ration for Market. Formulaes 27-38. Pages 13-16

CHAPTER III.
CONDITION POWDERS AND TONICS.

Condition Powders; Egg Foods; Poultry Tonics and Tissue Food. Formulaes 39-50. Pages 17-21

CHAPTER IV.
INSECTICIDES.

Insect Powders; Lice Liquids; Lice Killing Nest Eggs; Scab-mite Remedy; Scaly Leg Remedy. Formulaes 51-61. Pages 22-26

CHAPTER V.
ROUP CURES AND OTHER REMEDIES.

Miscellaneous Collection of Forty-Five Valuable Formulae in Alphabetical Order, Including Roup Cures; Canker Remedies; Catarrh Powders; Pills; Ointments; Gape Remedies; Cholera Remedy, and Other Valuable Prescriptions. Formulae 62-106. Pages 27-38

CHAPTER VI.
TREATMENT FOR POISONING.

Antidotes for Poisoning of Fowls Caused by Arsenic, Copper, Ergot, Lead, Phosphorus, Salt, Strong Lye, Quick-Lime and Soda Nitrate. Formulae 107-112. Pages 39-40

CHAPTER VII.
DISINFECTANTS.

Creolin, Formaldehyde and Sulphuric Acid Disinfectants. Kerosene Emulsion, Colored Whitewash and Whitewash for Spraying. Formulae 113-121. Pages 41-43

CHAPTER VIII.
EGG PRESERVATIVES.

Water Glass Method; Lime Water Brine and Southern Lime Method. Formulae 122-124. Pages 44-45

CHAPTER IX.
MISCELLANEOUS.

Cement for Floors; Curing Clover Hay; Hen Manure Fertilizers; Storage of Vegetables for Winter Feeding. Formulae 125-133. Pages 46-48

CHAPTER I.

BALANCED-RATION FOODS.

In this chapter will be found formulae for balanced-ration or ready-mixed poultry foods including scratch-grain mixtures, dry mash, moist-mash mixtures, chick foods, and home-prepared meat food. All of these recipes have been carefully tested by many years practical use and are entirely reliable and dependable. No reference is made to the chemical composition of foodstuffs and these balanced-ration mixtures are based on practical merit, not chemical composition. Anyone desiring to figure out the precise chemical contents of foodstuffs named will find complete analyses of all the common stock feeding materials in Farmer's Bulletin, No. 22, entitled "The Feeding of Farm Animals," published by U. S. Department of Agriculture. This bulletin can be obtained free on application to the Secretary of Agriculture.

In selecting a food ration the poultryman should be governed by the practical results obtained. Chemical analyses of food merely show the amount of water, fibre, protein, ash, fat, and non-nitrogenous matter the chemist is able to find therein. They do not indicate in any way the digestibility of the food. Some foodstuffs may be particularly rich in protein and at the same time that protein may be of little actual, practical value to the fowl because of difficulty in properly digesting it. Some foods rich in protein, notably beans and peas, also contain irritating or toxic elements that are actually injurious if such foods are fed continuously in considerable quantities. Heavy feeding of vegetable foods rich in protein is always much more likely to cause trouble and serious digestive disturbance than heavy feeding of animal protein foods like pure beef scrap, fresh meat, cut green bone, etc.

In mixing grain foods, the formulae for which are given in this chapter, the whole quantity of grain to be mixed should be dumped upon a smooth clean board floor. Here it should be mixed by turning over with scoop shovels until thoroughly blended. It can then be sacked or placed in bins.

As a general rule two dry grain feedings per day where regular feedings are given are sufficient. The daily allowance of dry grain per bird averages from 3 to 5 ounces per fowl, according to size and appetite. One quart of grain will weigh approximately one and one-half pounds and will contain about 16 fair-sized handfuls. One handful per adult fowl is a common allowance for one feeding of whole grain. In poultry feeding a handful of grain usually means half a gill. This quantity will weigh approximately 1.5 to 1.8 ounces. Fowls fed liberally on moist mashes will as a rule consume from 2 to 2.5 ounces per bird of dry ground grain daily in addition to the whole grain or scratch-grain fed.

Exercise or scratch-grain mixtures containing whole and cracked hard grain may be fed either in litter or from a food hopper. Dry mashes should be hopper fed. Moist mashes, where used, should be fed once daily usually, not oftener than five days a week, the other feedings being of whole grain or scratch-grain mixtures. Green food should be given daily, preferably just before noon, as much as the fowls will clean up before the night feeding. When meat is not fed in some other way, wholesome beef scrap should be kept before the fowls all the time in a food hopper. Grit, oyster shell, charcoal and granulated or cracked bone should be kept before poultry, young and old, at all times.

WHOLE GRAIN HOPPER-FED MIXTURES.

Spring and Fall Mixture.

1.—R.	Cracked Yellow Corn,	40 lbs. av.
	Whole Corn,	10 lbs. av.
	Hard Red or Amber Wheat,	30 lbs. av.
	Heavy Clipped White Oats,	10 lbs. av.
	Barley,	10 lbs. av.

Oats should be heavy white oats running from 38 to 42 lbs. to the bushel. Oats may be substituted for barley or vice versa, according to which can be purchased at lower price.

Winter Mixture.

2.—R.	Whole Corn,	30 lbs. av.
	Cracked Corn,	35 lbs. av.
	Wheat,	25 lbs. av.
	Oats,	10 lbs. av.

Five pounds of either barley or buckwheat may be substituted for five pounds of the oats, or both may be substituted for entire amount of oats, according to market prices and convenience.

Summer Mixture.

3.—R.	Cracked Corn,	30 lbs. av.
	Wheat,	40 lbs. av.
	Oats,	15 lbs. av.
	Barley,	15 lbs. av.

Ten pounds of kaffir corn may be substituted for five pounds each of oats and barley, if desired.

General Farm Mixture.

4.—R.	Sifted Cracked Corn,	800 lbs. av.
	Heavy Oats (with hulls on),	600 lbs. av.
	Good Hard Wheat,	600 lbs. av.

This formulae makes one ton of labor or scratch food which may be fed in litter or from a food hopper.

EXERCISE OR SCRATCH GRAIN FOODS.

The following grain mixtures are intended to be fed as labor or exercise foods in deep clean litter, but may be fed from a food hopper when more convenient with entirely satisfactory results.

Variety Grain Mixture.

5.—R.	Cracked Corn,	40 lbs. av.
	Whole Wheat,	20 lbs. av.
	Clean Wheat Screenings,	20 lbs. av.
	Kaffir Corn,	8 lbs. av.
	Oats,	6 lbs. av.
	Silverskin Buckwheat,	4 lbs. av.
	Sunflower Seed,	1 lb. av.
	Golden Millet,	½ lb. av.
	Whole Flaxseed,	¼ lb. av.
	Hemp Seed,	¼ lb. av.

High-Protein Exercise Food.

6.—R.	Wheat,	200 lbs. av.
	Heavy White Oats,	200 lbs. av.
	Sifted Cracked Corn (yellow preferred),	400 lbs. av.
	Barley,	100 lbs. av.
	Kaffir Corn,	60 lbs. av.
	Split or Broken Peas,	30 lbs. av.
	Hemp Seed,	4 lbs. av.
	Whole Flaxseed,	1 lb. av.
	Golden Millet,	5 lbs. av.

If there is any difficulty in getting kaffir corn, substitute 30 lbs. of silverskin buckwheat and add 30 lbs. to the wheat or oats. Cracked beans are sometimes used in place of peas.

DRY MASH MIXTURES.

These mixtures are intended for feeding adult fowls and are fed from a food hopper, being kept constantly before the birds. In addition to these, regular feedings of whole grain or scratch-grain mixtures are fed in litter once or twice daily.

Maine Dry Mash.

7.—R.
Wheat Bran,	20 lbs. av.
Yellow Corn Meal,	10 lbs. av.
Wheat Middlings,	10 lbs. av.
Gluten Meal,	10 lbs. av.
Linseed Meal,	10 lbs. av.
Fine-ground Beef Scrap,	10 lbs. av.

Above mixture is kept before the fowls all the time in an easy-access covered trough or hopper. In addition each 100 hens receive early in the morning 4 quarts screened cracked corn scattered in litter. A second feeding is given at ten o'clock consisting of 2 quarts whole wheat and 2 quarts oats. About 5 lbs. of cut clover are fed dry daily to each 100 birds. In winter mangels or other vegetable food are supplied in moderate quantity, as much as the birds will clean up readily. In summer time the flocks range on grass land. Dry, cracked bone, grit, charcoal, oyster shell and pure water are kept before the birds all the time. According to bulletin published by Maine Agricultural Experiment Station, the average amount of materials eaten by each hen during one year were as follows: Grain and meal mixture, 90 lbs.; oyster shell, 4 lbs.; grit, 2 lbs.; dry cracked bone, 2.4 lbs.; charcoal, 2.4 lbs.; clover, 10 lbs. Approximate cost of above yearly allowance for one hen, $1.45. Average egg production for one hen, one year, 144 eggs.

Dry Mash.

8.—R.
Wheat Bran,	400 lbs. av.
Yellow Corn Meal,	300 lbs. av.
Fancy Wheat Middlings,	200 lbs. av.
Ground Oats,	100 lbs. av.
Yellow Gluten Meal,	200 lbs. av.
Fine-ground Meat or Blood Meal,	100 lbs. av.
Clover or Alfalfa Meal,	200 lbs. av.

Where you can have your own grain ground to order at the local mill it is advisable to have the entire quantity of corn and oats ground together into a coarse meal known as corn and oat chop.

MOIST MASHES.

The following moist mashes are intended for breeding and laying stock fed either morning, noon or night at the convenience of the poultryman, once daily five or six days a week, the other regular feedings being of whole grain or of scratch-grain mixtures.

Cooked Vegetable Mash.

9.—R. Equal parts by measure corn meal, wheat middlings, bran, ground oats, and meat scrap or meat meal.

Add above mixture to one-fourth bucket of thoroughly cooked mashed vegetables,—beets, turnips, potatoes, carrots, apples, etc., and mix with sufficient hot water to make one bucketful of stiff, crumbly mash. To each bucket add one level tablespoonful of salt and one level teaspoonful of black pepper.

Winter Mash.

10.—R.
Ground Corn,	200 lbs. av.
Gluten Meal,	200 lbs. av.
Oat Middlings,	250 lbs. av.
Crushed Oats,	100 lbs. av.
Wheat Bran,	200 lbs. av.
Wheat Middlings,	50 lbs. av.

Where crushed oats and oat middlings are difficult to obtain, use 300 lbs. ground oats to take their place, and add 50 lbs. to the wheat middlings. This mash should be mixed with scalded cut clover, 10 per cent., and good beef scrap, 6 to 8 per cent. Clover should be scalded with water lightly seasoned with salt; cold water should be added to mix mash.

Stale Bread Mash.

11.—R.
Stale Bread,	20 lbs. av.
Corn Meal,	30 lbs. av.
Wheat Middlings,	10 lbs. av.
Wheat Bran,	20 lbs. av.
Cut Clover or Clover Meal,	5 lbs. av.
Ground Oats,	5 lbs. av.
Beef Scrap,	10 lbs. av.

Corn and Oat Chop Mash.

12.—R. Corn and Oat Chop,   200 lbs. av.
        Wheat Bran,   100 lbs. av.
        Beef Scrap,   10 lbs. av.

Corn and oat chop is made by grinding one bushel whole yellow corn with two bushels of oats into a rather coarse meal. Above mash should be mixed with cold skim milk. Cold water may be used where milk is not available, in which case the quantity of beef scrap should be doubled.

Laying Food Mash Mixture.

13.—R. Corn Meal (white or yellow),   40 lbs. av.
        Fancy Wheat Middlings,   20 lbs. av.
        Bran,   10 lbs. av.
        Ground Oats,   10 lbs. av.
        Fine-ground Beef Scrap or Meat Meal,   10 lbs. av.
        Fine-cut Clover or Clover Meal,   10 lbs. av.

Above mash mixture is mixed with cold water and fed as either the morning or night feed, once daily five or six days a week, giving as much as the birds will clean up in from fifteen to twenty minutes. One other grain feeding should be given daily of either whole grain,—corn, wheat, oats or buckwheat, or scratch-grain. On days on which mash is not fed scratch grain mixtures or whole grain are used in its place and cut green bone or meat scraps are fed.

FOOD MIXTURES FOR SMALL CHICKENS.

Chick Food, No. 1.

14.—R. Sifted Fine-cracked Yellow Corn
           or Corn Grits,   45 lbs. av.
        Cracked Hard Wheat,   20 lbs. av.
        Steel-cut Oats or C. Oatmeal,   20 lbs. av.
        Cracked Barley (hulls sifted out),   12 lbs. av.
        Granulated Willow Charcoal,   1 lb. av.
        Granulated Raw Bone (kiln dried),   1 lb. av.
        Chick-size Grit,   ½ lb. av.
        Golden Millet,   ½ lb. av.

Old-Fashioned Home-Ground Chick Food.

15.—R. Whole Yellow Corn,   one-half.
         Whole Wheat,   one-fourth.
         Hulled Oats,   one-eighth.
         Barley (with hulls on),   one-eighth.
         Parts given are by measure.

Mix all together and grind in grain mill or large-sized coffee mill to suitable size for chick feeding. Sift meal out by allowing crushed grain to run over an inclined screen of wire mosquito netting. Waste meal from this mixture may be added to mashes for adult fowls or fed dry to chicks.

Chick Food, No. 2.

16.—R. Corn Grits or Sifted Fine-
            cracked Corn,   50 lbs. av.
         Cracked or Steel-cut Wheat,   30 lbs. av.
         Cracked Barley (hulls sifted out), 10 lbs. av.
         Steel-cut Oats or C. Oatmeal,   8 lbs. av.
         Golden Millet,   1 lb. av.
         Granulated Raw bone (kiln dried),   1 lb. av.

Chick Food, No. 3.

17.—R. Sifted Fine Cracked Corn or Corn Grits,
                                              25 lbs. av.
         Steel-cut Oats or C. Oatmeal,   15 lbs. av.
         Cracked Peas,   5 lbs. av.
         Cracked Wheat,   40 lbs. av.
         Granulated Bone (kiln dried),   1 lb. av.
         Fine-cracked Kaffir Corn,   10 lbs av.
         Cracked Buckwheat (hulls sifted out), 1 lb. av.
         Golden Millet,   2 lbs. av.
         Chick Grit,   1 lb. av.

Chick Food, No. 4.

18.—R. Sifted Fine-cracked Corn
             or Corn Grits,   350 lbs. av.
         Cracked Kaffir Corn,   60 lbs. av.
         Fine-cracked Peas,   40 lbs. av.
         Steel-cut or Cracked Wheat,   200 lbs. av.
         Steel-cut Oats,   100 lbs. av.
         Clean, Fine Wheat Screenings,   200 lbs. av.
         Golden Millet,   25 lbs. av.
         Cracked Buckwheat (hulls sifted out),
                                              25 lbs. av.

Growing Food.

19.—R.	Cracked Corn,	40 lbs. av.
	Whole Wheat,	30 lbs. av.
	Kaffir Corn,	10 lbs. av.
	Clean Wheat Screenings,	10 lbs. av.
	Hulled Oats,	10 lbs. av.

The above is intended for food for weaned chicks to be fed in the place of chick food after they are from six to eight weeks old.

Four Grain Mixture.

20.—R.	Whole Corn,	50 lbs. av.
	Whole Wheat,	18 lbs. av.
	Heavy White Oats,	16 lbs. av.
	Heavy Barley,	16 lbs. av.

Mix together and have ground to flour fineness. This grain mixture may be kept before growing chicks all the time in a food hopper. In addition they should have beef scrap, whole grain mixtures, and an abundant supply of green food. Give the regular allowance of granulated bone, charcoal, grit, oyster shell and pure water.

Forcing Food Mash.

21.—R.	Yellow Corn Meal,	50 lbs. av.
	White Wheat Middlings or Low-Grade Flour,	10 lbs. av.
	Meat or Blood Meal,	20 lbs. av.
	Bran,	10 lbs. av.
	Clover Meal,	10 lbs. av.

To be made into a crumbly moist mash with skim milk or cold water and fed to growing chicks after six weeks old in addition to cracked corn, wheat and growing food. May be fed dry if desired.

MISCELLANEOUS FOOD MIXTURES.

Moist Mash for Young and Old Stock.

22.—R. Equal parts by measure wheat bran and oat feed or ground oats. Mix with sufficient fine-cracked corn to make the mash dry and crumbly. Add 20 per cent. beef scrap and from 40 to 50 per cent. cooked cut clover or cooked vegetables. Thoroughly cooked fresh waste fish may be used in place of the beef scrap when obtainable. This mash is fed to laying fowls and to growing chicks over one month old. In addition to the mash, feed once daily a mixture of dry grains, three parts corn and one part of either oats, wheat or barley.

English Fattening Ration.

To be fed twice daily from a trough to birds confined in fattening coops. Makes white-fleshed poultry desired in English market.

23.—R. For the first week give wheat bran, ground oats (coarse hulls sifted out), equal parts by measure, mixed into rather sloppy gruel with either skim milk or pure sour milk.

For the second week give ground oats alone mixed with skim milk or sour milk. Milk used should be either always sweet or always sour.

Third week to finish, add one tablespoonful of mutton fat for each bird to the ground oats and milk gruel. The fowls should be given all that they will clean up at one meal. Feed twice a day.

White-Flesh Fattening Mixture for Crate Fattening.

24.—R. Ground Oats (hulls sifted out), 10 lbs. av.
Ground Buckwheat (coarse hulls sifted out), 10 lbs. av.
Ground White Corn, 5 lbs. av.

Mix into rather sloppy gruel with skim milk and feed from troughs two or three times daily to birds confined in fattening crates.

Yellow-Flesh Fattening Mixture.

25.—R. Fine Yellow Corn Meal, 10 lbs. av.
Ground Oats (hulls sifted out), 5 lbs. av.
Ground Buckwheat (hulls sifted out), 5 lbs. av.
Cotton Seed Meal, 1 lb. av.

Mix into thick, rather sloppy gruel with skim milk and feed from troughs to crate fatten.

Any mash mixture or dry grain mixture combined with meat food, if fed heavily to fowls that are closely confined, will serve as a fattening food.

Scrapple or Home-Made Meat Food.

26.—R. All sweet, fresh, waste meat, blood and bones may be made available for poultry food including clean feet and hooves of beef, sheep or hogs. Grind all to fineness of coarse-cracked corn in a bone cutter. Place in a large kettle, season with salt, black pepper and sage leaves, cover well with water. Boil thoroughly for five or six hours. Skim off and save excess of fat. When meat and bones have been thoroughly boiled for at least five hours, begin to stir in yellow corn meal, while boiling, until mixture makes a thick dough. Cook corn meal thoroughly, stirring frequently to prevent burning. When cooked remove into stone jars or clean wooden butter tubs, and pack down well. Over the top pour a thick coat of hot melted fat which was skimmed from the boiling mixture. This home-made meat food will keep sweet several weeks if stored in a cool, dry place, even in summer weather. Feed all the fowls will clean up quickly at one meal four times a week. This is a first class substitute for beef scrap.

CHAPTER II.

DUCK RATIONS.

On many duck ranches the ducks are fed exclusively on moist mash food, on others on a combination of moist mash and dry grains. In this chapter are given the most satisfactory rations for young and old ducks.

New Jersey Duck Mash.

27.—R.	Wheat Bran,	200 lbs. av.
	Corn Meal,	100 lbs. av.
	Ground Oats,	100 lbs. av.
	Low Grade Flour,	75 lbs. av.
	Beef Scrap,	75 lbs. av.

With every five pails of this mixture are used two pails of dry cut clover (loosely packed in pail). The clover is scalded and lightly salted before mixing with the grain. The grain mixture is added to the wet clover in the warm clover tea from scalding and mixed into a stiff, crumbly mash. Mash is fed either end of the food trough. Fine grit and fine crushed oyster shell are kept before the ducks all the time.

Summer Ration for Breeding Ducks.

28.—R.	Heavy Wheat Bran,	3 pecks.
	Low Grade Flour,	1 peck.
	Corn Meal,	1 peck.
	Best Beef Scrap,	3 lbs. av.
	Fine Grit or Sand,	1 lb. av.

Mix with cold water. Feed twice daily. This ration is for ducks kept on wide green range or pasture in summer season and intended for breeding stock.

Laying Ration for Breeding Ducks.

29.—R.
Wheat Bran,	1 bushel.
Corn Meal,	1 bushel.
Beef Scrap,	10 lbs. av.
Low Grade Flour,	1½ pecks.
Boiled Vegetables, turnips, beets, or potatoes,	1 peck.
Scalded Cut Clover or Clover Rowen, Green Rye or Cabbage,	1½ pecks.
Grit or Sand,	3 pounds av.

Above mash should be mixed with cold water and fed night and morning. At noon give light feeding of cracked corn and whole oats. Keep fine grit, fine ground oyster shell and pure water always before the birds.

Ration for Breeding and Laying Ducks.

30.—R.
| Corn Meal, Shorts or Mixed Feed, | equal parts. |
| Best Beef Scrap, | 10 per cent. |

Add a little clean sharp sand.

To above mixture is added in season fresh, finely-cut green food or cured green food in the proportion of four buckets of green stuff to each fourteen buckets of grain. The whole is mixed dry and then made into a crumbly mash with cold water. Sufficient low grade flour should be added to just stick the mash together, so that it will cling in shape when made into a ball with the hands. Feed this mash morning and night, all they will clean up in fifteen to twenty minutes. At noon give a light feeding of whole corn.

Ration for Ducklings Under Two Weeks.

31.—R. Give a mash of equal parts by measure corn meal and shorts, with a little sharp sand. Feed five times a day. Give no scrap or green food until two weeks old. Shorts is a term for a by-product of wheat commonly called in some localities mixed feed, being a mixture of bran, re-ground bran and middlings, in presumably equal parts.

Ration for Ducklings Two Weeks to Eight Weeks Old.

32.—R. Equal parts by measure corn meal and shorts, with 10 per cent. best beef scrap and a little clean, sharp sand. Add, according to season and supply, 10 to 15 per cent. fresh, finely-cut green clover or rye, or scalded fine-cut clover (cured). Feed four times a day until five weeks old, then three times a day until eight weeks old.

Finishing-Off Mash for Market Ducklings.

Two weeks before the birds are killed feed the following ration, omitting all green food.

33.—R.
Corn Meal,	2 bushels.
Ground Oats,	1 bushel.
Beef Scraps,	15 pounds av.
Low Grade Flour,	2 pecks.

Mix with cold water. Feed three times daily.

Ducklings Under Four Days Old.

34.—R.
Wheat Bran,	1 bushel.
Corn Meal,	1 peck.
Low Grade Flour,	1 peck.
Fine Grit,	2 pounds av.

Mix into crumbly mash with cold water and feed four times a day all they will clean up in twenty minutes.

Ducklings Four Days to Four Weeks Old.

35.—R.
Wheat Bran,	1 bushel.
Corn Meal,	1 peck.
Low Grade Flour,	1 peck.
Fine-ground Beef Scrap,	3 pounds av.
Fine Grit,	1½ pounds av.

Beef scrap should be scalded before mixing with grain. Mix mash with cold water. Feed four times a day all they will clean up. Give cut green clover, fresh cut rye, or cabbage freely.

Ducklings Four to Six Weeks Old.

36.—R.
Wheat Bran,	3 pecks.
Corn Meal,	1 peck.
Low Grade Flour,	1 peck.
Fine Grit,	1 pound av.
Fine-ground Oyster Shells,	½ pound av.
Scalded Beef Scrap,	3 pounds av.

Mix in 10 per cent. fine cut, fresh green food. Mix mash with cold water; feed four times a day all they will clean up.

Ducklings Six to Eight Weeks Old.

37.—R. Wheat Bran, 1 bushel.
Corn Meal, 1 bushel.
Low Grade Flour, 10 quarts.
Beef Scrap, 12 pounds av.
Fine-cut Fresh Green Food, 1 peck.
Grit, 3 pounds av.

Mix into crumbly mash with cold water. Feed all they will clean up three times a day. Keep oyster shell before them.

Ducklings Eight to Eleven Weeks Old.

38.—R. Corn Meal, 1 bushel.
Wheat Bran, 2 pecks.
Low Grade Flour, 2 pecks.
Beef Scrap, 10 pounds av.
Grit, 2 pounds av.

Mix into crumbly mash with cold water. Keep oyster shell always before the birds. Feed green food with the mash at first, then less freely until within ten days to two weeks of market time. Then omit green food altogether.

CHAPTER III.

CONDITION POWDERS AND TONICS.

The formulae of condition powders and tonics found in this chapter have been selected from a large collection of recipes for preparations of this sort, and are considered the best of their kind. As a rule I do not recommend the use of tonics or condition powders for healthy fowls, but many practical poultrymen disagree with me on this subject, and like to feed condimental foods or stimulating tonics to their birds, believing that by so doing they are able to keep them in better condition and get better results. Most of the so-called condition powders, egg makers, tonics and the like, can be cheaply prepared by the poultryman at a great saving over the cost of the commercial article, with the further advantage that you know of just what materials they are made up.

Condition Powder, No. 1.

39.—R. Capsicum (pure red pepper), 1 lb. av.
Sodium Chloride (table salt), 2 lbs. av.
Iron Sulphate (copperas, powdered fresh), 2 lbs. av.
Sodium Bicarbonate (baking soda), 2 lbs. av.
Mustard (fine ground), 2 lbs. av.
Magnesium Oxide (magnesia), 2 lbs. av.
Sulphur (pure flowers of sulphur), 2 lbs. av.
Calcium Carbonate (powdered chalk), 5 lbs. av.
Oyster Shell (fine ground powder), 5 lbs. av.
Raw Bone (kiln-dried, fine ground), 5 lbs. av.
Charcoal (pure willow, fine-ground), 2 lbs. av.
Fine White Sand (silica), 10 lbs. av.
Blood Meal (kiln-dried), 20 lbs. av.
Fenugreek, 20 lbs. av.
Oil Cake (fine ground old process linseed meal, 20 lbs. av.

All of above ingredients should be reduced to a fine powder or meal and thoroughly mixed. Some of the ingredients, as will be noted, can be bought more cheaply from a feed dealer or from a grocery, while the balance should be obtained from your druggist. This powder is recommended for improving the condition of the stock, growing lustrous plumage and increasing the egg production.

DIRECTIONS:—Mix one heaping tablespoonful of above powder with sufficient mash food for 20 hens, or with about two quarts of a dry ground grain mixture intended for making the mash. Add cooked vegetables or scalded cut clover to mash to give bulk.

Condition Powder, No. 2, or Egg Food.

40.—R. Potassium Nitrate (saltpetre), 12 ounces av.
Iron Sulphate (copperas
  powdered fresh), 20 ounces av.
Table Salt (sodium chloride), 12 ounces av.
Powdered Charcoal, 10 ounces av.
Flowers of Sulphur, 18 ounces av.
Pulverized Oil Cake (old
  process linseed meal), 30 ounces av.
Fenugreek, 28 ounces av.
Blood Meal (kiln-dried), 70 ounces av.

Reduce each of above ingredients to a powder and then thoroughly mix all together.

DIRECTIONS:—Mix one heaping tablespoonful with sufficient mash food for 20 hens, or the same quantity may be mixed with about two quarts of dry ground grain intended for making the mash.

Condition Powder, No. 3.

41.—R. Fine-ground Raw Bone (kiln-dried), 50 lbs. av.
Powdered Fenugreek, 10 lbs. av.
Pulverized Oyster Shell, 10 lbs. av.
Powdered Calcium Phosphate, 10 lbs. av.
Black Pepper, 10 lbs. av.
Powdered Gentian, 8 lbs. av.
Venetian Red (red iron oxide), 2 lbs. av.
Reduce all to powder and thoroughly mix.

DIRECTIONS.—Use one heaping teaspoonful to each quart of moist mash.

Condition Powder, No. 4.

42.—R. Pulverized Oyster Shell, 58 ounces av.
Capsicum (powdered red pepper), 2 ounces av.
Powdered Ginger, 10 ounces av.
Iron Sulphate, 6 ounces av.
Flowers of Sulphur, 4 ounces av.
Powdered Gentian, 20 ounces av.
Reduce all to powder and thoroughly mix.

DIRECTIONS:—Use one rounded tablespoonful with sufficient mash food for 20 hens, or use the same quantity in two quarts of dry ground grain intended for mixing moist mash.

Condition Powder, No. 5.

43.—R.　Sodium Sulphate,　　　　　　30 ounces av.
　　　　Sulphur,　　　　　　　　　　14 ounces av.
　　　　Powdered Fenugreek,　　　　 10 ounces av.
　　　　Powdered Gentian,　　　　　　9 ounces av.
　　　　Powdered Ginger,　　　　　　 9 ounces av.
　　　　Powdered Black Antimony,　　 8 ounces av.
　　　　Fine-ground Raw Bone Meal (kiln
　　　　　dried),　　　　　　　　　 10 ounces av.
　　　　Blood Meal (kiln-dried),　　 10 ounces av.

Reduce all to fine powder and thoroughly mix.

DIRECTIONS:—One teaspoonful to each quart of mash food.

Condition Powder, No. 6.

44.—R.　Powdered Gentian Root,　　　5 ounces av.
　　　　Powdered Calamus Root (sweet
　　　　　flag),　　　　　　　　　　 5 ounces av.
　　　　Powdered Buckbean Leaves (meny-
　　　　　anthes),　　　　　　　　　 3 ounces av.
　　　　Powdered Sabbatia (herb),　　6 ounces av.
　　　　Powdered Poke Berries (phytolacca
　　　　　decandra),　　　　　　　　 8 ounces av.
　　　　Powdered Juniper Berries,　　8 ounces av.
　　　　Powdered Leaves of Bitter Orange,
　　　　　　　　　　　　　　　　　　 5 ounces av.
　　　　Powdered Wormwood (herb),　　5 ounces av.
　　　　Powdered Cinnamon,　　　　　 2 ounces av.
　　　　Powdered Ginger,　　　　　　 2 ounces av.
　　　　Powdered Mustard Seed,　　　 1 ounce av.

All of above can be purchased of a wholesale druggist, should be fine powdered and thoroughly mixed.

DIRECTIONS.—Use one teaspoonful to two quarts of dry ground grain used for mixing the mash, and add one rounded teaspoonful of table salt. This powder is claimed to be very beneficial for breeding stock and laying hens. Corrects tendency to internal fat. Valuable for molting fowls. See formula R. No. 50.

Condition Powder, No. 7.

45.—R.　Powdered Fenugreek,　　　　 8 ounces av.
　　　　Potassium Nitrate,　　　　　 4 ounces av.
　　　　Powdered Gentian,　　　　　　1 pound av.
　　　　Powdered Ginger,　　　　　　 4 ounces av.
　　　　Flowers of Sulphur,　　　　　2 ounces av.
　　　　Sulphate of Iron,　　　　　　2 ounces av.
　　　　Black Antimony,　　　　　　　2 ounces av.
　　　　Flaxseed Meal,　　　　　　　 8 ounces av.

Reduce all to fine powder and thoroughly mix. Use one teaspoonful to a quart of mash food for adult fowls.

Egg Food or Forcer.

46.—R.
Calcium Phosphate,	12 ounces av.
Sodium Chloride (table salt),	8 ounces av.
Calcium Carbonate,	6 ounces av.
Powdered Ginger,	4 ounces av.
Flowers of Sulphur,	2 ounces av.
Potassium Nitrate,	1 ounce av.
Powdered Cantharides,	1 dram av.

Reduce all to fine powder and thoroughly mix. Use one teaspoonful in one quart of mash food for laying fowls.

English Poultry Tonic.

47.—R.
Sugar,	1 pound av.
Sulphate of Iron,	2 drams av.
Sulphate of Magnesia (Epsom Salts),	2 ounces av.
Sulphate of Sodium (Glauber's salts),	4 ounces av.
Chloride of Sodium (table salt),	½ ounce av.
Water sufficient to make one quart.	

The sugar should first be placed in a pan and boiled and burnt until it makes a thick dark brown syrup. The other ingredients in finely powdered form should be dissolved in water and added to burnt sugar syrup. One tablespoonful of above tonic may be added to each quart of drinking water to build up debilitated fowls and stimulate egg production.

Home Iron Tonic.

48.—R.
Compound Tincture Gentian,	2 teaspoonfuls.
Tincture Iron Chloride,	2 teaspoonfuls.
Lime Water,	4 tablespoonfuls.
Eggs,	2.
Cod Liver Oil,	1 gill.

Mix, place in bottle and shake thoroughly. As a tonic and conditioner for show birds or laying stock that are out of condition add two tablespoonfuls of above tonic to one quart of moist mash food. For individual treatment an adult bird may be given one teaspoonful twice daily.

Tonic for Show Birds.

49.—R. Hensel's Physiological Tonicum.

An iron preparation obtainable at any reliable Homoeopathic Pharmacy. It is an exceptionally fine iron tonic for debilitated show birds to put them in good show condition. This remedy is expensive and is only recommended for valuable show birds. It is particularly useful where the bird's comb turns dark and specimen loses appetite; is dumpish, and generally out of condition. Use one teaspoonful in one gallon of drinking water, allowing birds no other drink. In severe cases use only one-half gallon of water to one teaspoonful of tonic.

Tonic Powder or Tissue Food.

50—.R.
Bone Meal (kiln-dried),	5 pounds av.
Pulverized Oyster Shell (calcium carbonate),	1 pound av.
Potassium Sulphate,	1 pound av.
Sodium Chloride (table salt),	2 pounds av.
Sodium Phosphate,	4 ounces av.
Sodium Acetate,	4 ounces av.
Sodium Sulphate (Glauber's salts),	4 ounces av.
Calcium Fluoride,	1 ounce av.
Magnesium Phosphate,	10 ounces av.
Ferric Trioxide (red oxide of iron),	10 ounces av.
Ammonium Sulphate,	14 ounces av.
Manganese Dioxide (black oxide of manganese),	½ ounce av.
Silicic Acid,	½ ounce av.

Reduce all to powder and thoroughly mix. One pound of above mixture may be added to 100 lbs. of dry-ground-grain, mash-mixture, to be used as moist or dry mash. For laying fowls, it is considered beneficial to also add one-half pound of condition powder No. 6 (R. No. 44), to the same quantity of dry ground grain. For chicks two to six months old, use one half of above quantities per hundred pounds of grain. Particularly useful where fowls are kept in rather close confinement on so-called "intensive plan" of poultry culture. Prevents leg weakness in young or old stock, growing light and debility. Useful during molt to insure quick growth of plumage and to build up and strengthen fowls body. Considered one of the best general conditioners when used with Condition Powder No. 6.

CHAPTER IV.

INSECTICIDES.

In this chapter will be found reliable formulae for lice and mite killing mixtures in both powder and liquid form. While the several formulae which follow give entire satisfaction to those who use them, I wish to state emphatically that there is only one best insect powder for use on fowls and chicks. It costs a little more than some of the mixtures, the formulae for which appear below, but it is by far the most satisfactory insect powder I have ever used. This powder is simply the pure fresh-ground flowers of Persian Pellitory or Pyrethrum, which is sold under the name Persian insect or Dalmation powder. It should be made of the partly-opened flower heads, *fresh ground*. If old or adulterated it will not prove effective. If pure and fresh it is the most reliable insecticide the poultryman can obtain. The pure powder should be thoroughly dusted into the fowl's plumage clear down to the skin. It will not injure newly hatched chicks or poultry of any age, and may be used freely. Thoroughly applied the results will prove effective and lasting, the fowls remaining practically free from lice for a period of three months or more, as I have proved many times by actual test, except in the very warmest summer weather when lice apparently breed with remarkable rapidity. Even during this hot season two thorough dustings with pure Pryethrum powder, ten days apart, will insure practical freedom from lice throughout the summer months, if applied early in June.

Its usual cost is about 30 cents per pound for the pure, fresh article, but can be bought cheaper in 5, 10 or 25 pound lots. Pure, fresh-ground Persian insect flowers or Pyrethrum powder can be obtained through the Eastern Drug Co., Boston, Mass., who are importers of the genuine flowers. Much of the so-called Dalmation insect powder that is on the market is adulterated with flour, chalk, dust or other material to give bulk.

Insect Powder, No. 1.

51.—R. Tobacco Dust, 8 ounces av.
Pulverized Naphtalene (tar camphor), 1 ounce av.
Dalmation Insect Powder, 5 ounces av.
Sulphur, 4 ounces av.
Powdered Chalk or Marble Dust sufficient to make 2 pounds av.

All should be reduced to fine powder and thoroughly mixed. The last ingredient is used only to increase the bulk and give weight to the mixture. It may be applied in the same manner as any lice powder. Tobacco dust can usually be purchased in quantity through supply houses or from any dealer in tobacco products.

Insect Powder, No. 2.

52.—R. Tobacco Dust,   40 lbs. av.
        Chalk or Marble Dust,   5 lbs. av.
        Powdered Naphtalene,   5 lbs. av.

Each ingredient should be reduced to a fine powder. The powdered chalk and pulverized naphtalene flakes should be well rubbed together and afterwards thoroughly mixed with the tobacco dust. This is used as a lice or dusting powder for young and old fowls.

Insect Powder, No. 3.

53.—R. To a peck of freshly air-slaked lime add half an ounce of 90 per cent. carbolic acid mixed with one pint of water and stir thoroughly. Let stand in a covered box for two or three days. Reduce to a fine powder. To this fine powdered carbolized lime add an equal quantity by bulk of fine tobacco dust and thoroughly mix. To 16 ounces of this may be added 4 ounces of flowers of sulphur if desired. Use the same as any lice powder.

Insect Powder, No. 4.

54.—R. To half a peck of finely pulverized coal ashes add 8 fluid ounces (one half pint), of Lice Liquid No. 1 (R. No. 56). Stir well and let stand in covered box for two or three days. When dry add half a peck of fine tobacco dust and thoroughly mix. This makes an effective dusting powder. Tobacco dust usually costs from 2½ to 5 cents a pound. It should be borne in mind that to be effective any insect powder must be thoroughly applied, worked well into the fowls feathers down to the skin.

Insect Powder, No. 5.

55.—R. Persian Pellitory (pure Pyrethrum
        powder),   25 lbs. av.
       Tobacco Dust (pure fine-ground),   60 lbs. av.
       Naphtalene (fine powder sifted),   10 lbs. av.
       Marble Dust or Powdered Chalk,   5 lbs. av.

Reduce to a fine powder and thoroughly mix. Use as a dusting powder, working it well into the feathers down to the skin. Use every five or six weeks as a preventive.

CAUTION.—This powder is strong and when young chicks are treated they should not be confined immediately afterward in boxes or tight coops where fresh air is scarce.

Lice Liquid, No. 1.

56.—R. Crude Naphtalene Flakes,   18 ounces av.
        Kerosene (coal oil),   1 gallon.
        Creolin,   3 fluid ounces.

Dissolve the naphtalene in the kerosene, stirring well. Let the mixture stand two or three days. Then draw off the clear liquid and mix it with the creolin. Keep in tight tin can with screw cap or wooden stopper. This lice liquid or liquid lice killer is a very effective and satisfactory one. It may be used for painting the dropboards and roost poles to get rid of red mites, also for painting the bottoms of the nests for the same purpose. The vessel containing this liquid killer should not be exposed to a freezing temperature. Keep it at a temperature of about 60 degrees to 70 degrees F. or warm it to that temperature by letting it stand in a warm room before using.

DIRECTIONS.—To rid a hen of lice by means of this liquid killer, paint the inside bottom of a box or barrel with the liquid, place the hen therein and cover with a piece of coarse burlap or a basket, something that will partially confine the fumes and at the same time allow the fowl a little pure air. If the weather is very cold the box should be put in a warm place so that the fumes of the liquid may be readily given off. The fowl may remain thus cooped up in the box for twenty minutes to half an hour, when she should be taken out and allowed to have a run in the open air. A number of fowls may be treated at one time. It is not dangerous for adult birds to remain in the box several hours provided they are not covered in too tightly. Care should be used in treating young chickens, as they are sometimes susceptible to the fumes. Do not get the floor of the box so wet with the liquid that you saturate the bird's plumage. Remember that kerosene confined close to the fowl's body is liable to produce a blister. Sometimes fowls have been seriously blistered by careless use of kerosene preparations. The only inconvenience noted from proper use of this remedy is a slight looseness of the bowels in the treated birds, which usually wears off in 24 hours. Red mites will not venture on roosts that are saturated with this liquid once or twice a month. Sawdust slightly moistened with this Lice Liquid No. 1 is excellent for use in nests to keep away lice and mites. It should be scattered on the bottom of the nest and covered with hay or straw. This lice liquid can be produced at a cost not to exceed 35 cents a gallon.

Lice Liquid, No. 2.

57.—R. Crude Carbolic Acid (60 per cent.), 16 ounces.
Crude Naphtalene, 6 ounces.

Mix thoroughly. It may be used in the same manner as any liquid lice paint. The *crude* carbolic acid is a dark, oily liquid, a mixture of cresol and phenol, and should not be confused with the carbolic acid ordinarily purchased at drug stores. Your druggist can obtain it for you and also the naphtalene through his wholesaler. It can be used in the same manner as other liquid lice killers. Do not apply to roosts late in afternoon, as fowls roosting on perches wet with this liquid may blister or burn feet and breasts.

Lice Liquid, No. 3.

58.—R. Cresol, 1 pint.
Powdered Naphtalene Flakes, 12 ounces av.
Kerosene (coal oil), 3 quarts.

Dissolve naphtalene in kerosene and add cresol. Use in same manner as Lice Liquid, No. 1.

Lice Killing Nest Eggs.

59.—R. A very satisfactory lice killing nest egg may be made by simply melting crude naphtalene flakes and moulding them in the shape of an egg. An easy method is to blow out the contents of an egg by making a small hole in each end. After you have the egg shell clear and dry, plug up the hole in one end with a little plaster of Paris. The opening in the opposite end may be enlarged sufficiently to admit the tip of a very small pointed tin funnel. Melt your naphtalene in an ordinary double cooker, putting the crude flakes in the upper vessel, keeping water boiling in the lower vessel in which the upper one sits. Naphtalene melts at a temperature below that of boiling water. Pour the melted naphtalene slowly and carefully into the egg shell mould through funnel; the shell should rest in a pan of hot moist sand. After your lice killing nest egg is moulded in the natural egg shell, allow it to thoroughly cool. It may then be placed in the nest without removing the shell

Scab-Mites on Body and Feathers.

Collections of crusts about root of feathers; feathers break off close to skin.

60.—R. Naphtalene (powdered),   75 grains Troy.
        Sweet Oil,   150 drops.
        Flowers of Sulphur,   1 ounce Troy.
        Soft Soap,   1 ounce Troy.
        Lard or Wool fat,   1 ounce Troy.

Mix thoroughly. For scabies of body and plumage or scaly leg, apply to affected parts daily, rubbing in well.

Scaly Leg.

61.—R. Naphtalene Flakes,   4 ounces av.
        Kerosene,   1 quart.

Mix. For scaly leg, apply above to affected parts or mix in open quart can and dip fowl's legs in mixture. Be careful not to get above hocks on soft parts, for kerosene will blister soft flesh. Apply in daytime three times a week if necessary, and let fowl run after applying.

CHAPTER V.

ROUP CURES AND OTHER REMEDIES.

In this chapter are given a collection of miscellaneous formulæ, including some of the most reliable remedies for poultry diseases. As a rule these can be prepared very economically by the poultryman himself or by his druggist. All prescriptions in this chapter are given in apothecary's weights and measures, except where otherwise specified.

Abrasions.

62.—R. Powdered Boracic Acid,   1 ounce.
        Salicylic Acid,   3 grains.

Mix and use as a dusting powder, on raw, scratched or abraded surface, after first thoroughly cleansing with yellow soap and warm water.

Aconite, Bryonia and Spongia Mixture.

63.—R. Mother Tincture of Aconite (from
            whole plant),   10 drops.
        Mother Tincture of Bryonia Alba
            (from fresh root),   10 drops.
        Mother Tincture of Spongia (from
            roasted sponge),   10 drops.
        Alcohol sufficient to make one fluid ounce.

Mix thoroughly. Use one teaspoonful of this mixture in each quart of drinking water, allowing the birds no other drink. This is a very useful remedy in sudden colds, snuffles, the early stages of roup, catarrh, and all diseases in which frothy eyes and running at the nose are common symptoms. In many cases it will be found effective where one side of face is swollen and eye closed. These "mother tinctures" should be obtained from a Homœopathic physician or pharmacy.

Alum Water.

64.—R. Aluminum and Potassium Sulphate
(alum U. S. P.), 15 grains.
Water, 1 fluid ounce.

Mix. For adult fowls affected with cholera or diarrhœa give one to three teaspoonfuls of the alum water morning and night. Add one teaspoonful of above mixture to a pint of drinking water for little chicks affected with diarrhœa, and allow no other drink.

Baldness or Favus.

65.—R. Corrosive Sublimate (mercury
bichloride), 20 grains.
Glycerine, 4 fluid drachms.
Alcohol sufficient to make one pint.

Mix. CAUTION.—This remedy is POISON if taken internally and should be so labeled. It is useful as a wash or lotion in cases of baldness and favus in fowls and may be applied daily to the bald spots.

Blackhead Mixture.

66.—R. Sulphur, 10 grains.
Sulphate of Iron, 1 grain.
Sulphate of Quinine, 1 grain.

Mix above and give at one dose to a bird weighing 6 lbs. or over. This remedy is recommended for blackhead in turkeys.

Bruises.

67.—R. Spirits of Camphor, 1 fluid ounce.
Tincture of Arnica, 2 fluid ounces.
Distilled Extract Witch Hazel
sufficient to make one pint.

Mix and apply cotton or lint saturated with above mixture to all bruises to relieve pain and avoid discoloration.

Burns or Blisters.

68.—R. Linseed Oil, 7 fluid ounces.
Lime water (freshly prepared), 8 fluid ounces.

Mix and apply freely to burns and scalds or to blisters caused by careless use of kerosene.

Canker Remedy, No. 1.

69.—R.   Powdered Permanganate of
             Potassium,   1 grain.
          Powdered Sugar of Milk,   1 ounce.

Reduce both to fine powder and thoroughly mix. Do not use cane sugar. Blow this powder through a straw directly onto the canker ulcers or cheesy patches in the mouth and throat of affected birds. Also useful where ulcers, sores or cheesy patches are found about the vent. When it comes in contact with moist mucous surfaces the powder turns a purplish pink. Apply two or three times daily at first, then less often as the case improves.

Canker Remedy, No. 2.

70.—R.   Loeffler Solution (P. D. & Co.),   1 ounce.

Above prescription will be filled by any reliable druggist and is a preparation manufactured by Park, Davis & Co., of Detroit, Michigan, for the treatment of diphtheria in human beings. It is recommended here for valuable show birds affected with canker and is a reliable, effective remedy that can be depended upon, when properly applied, to result in a prompt cure. The affected parts should first be rubbed thoroughly with a bit of dry absorbent cotton. After this has been done saturate an absorbent cotton swab (made by twisting a bit of absorbent cotton around a sharp stick), with the Loeffler's Solution. Press this moist swab firmly against the canker patch or cheesy mass and allow it to remain for about ten seconds. Repeat this application immediately and thereafter three times a day until a cure is effected. If there is much accumulation of cheesy matter that will come away easily this should be removed before making the application.

Canker Remedy, No. 3.

71.—R.   Pulverized Camphor,   1 ounce.
          Powdered Boracic Acid,   1 ounce.
          Powdered Subnitrate of Bismuth,   1 ounce.

Mix and rub well together. For canker above powder should be blown into nostrils and throat through a straw or glass tube, or from a powder blower. Useful in canker and chronic catarrhal colds.

Catarrh Powder.

72.—R. Ginger, 4 drachms.
Gentian Root, 4 drachms.
Iron Sulphate, 2 drachms.
Sodium Hyposulphite, 1 drachm.
Sodium Salicylate, 1 drachm.

Reduce all to fine powder and mix. Use one level teaspoonful in sufficient moist mash for 15 adult fowls. Dose,—for a 5 to 6 lb. bird, from 3 to 4 grains once a day. Useful in all catarrhal colds.

Catarrh or Roupy Colds.

73.—R. Creolin-Pearson, 30 drops.
Camphor Water, 4 fluid ounces.
Glycerine, 1 fluid ounce.
Water sufficient to make one pint.

MIX.—For catarrh or roupy colds use two or three times daily as a spray or to swab out nostrils and mouth. Also useful as wash for sores in chicken pox.

Chicken Pox.

74.—R. Silver Nitrate, 2 grains.
Water, 1 ounce.

Mix and keep in brown glass bottle. For chicken pox apply to pox sores after removing the scabs.

Constipation.

75.—R. Fluid Extract Cascara, 1 fluid ounce.
Fluid Extract Licorice, 2 fluid drachms.

MIX.—For constipation in adult fowls give one teaspoonful at night. For chicks 5 to 20 drops according to age.

Contagious Catarrh.

76.—R. Olive Oil or Sweet Oil, 4 fluid ounces.
Finely-powdered Iodoform, 1 drachm.

Mix well and use as a spray in atomizer to spray throat and nostrils in cases of contagious catarrh, chronic catarrh or chronic roup. May also be applied with a feather if more convenient.

Corns.

Hard, horny hardening of skin of ball of foot.

77.—R.　Extract Cannabis Indica,　　　3 grains.
　　　　Salicylic Acid,　　　　　　　　30 grains.
　　　　Oil of Turpentine,　　　　　　15 drops.
　　　　Glacial Acetic Acid,　　　　　6 drops.
　　　　Flexible Collodion, sufficient
　　　　　to make　　　　　　　　　　5 fluid drachms.

Mix.—Bottle should be kept well corked. For corns apply with small brush to the corn to give a thin coating, and repeat each night until all drops off, bringing away the corn. Keep fowl confined where it cannot pick at foot.

Creolin Ointment.

78.—R. To one cupful of melted lard add one teaspoonful of pure creolin. Stir until cool. This is an effective application in chicken pox, scabies, scaly leg and similiar troubles. In chicken pox or where there are black scabs or crusts on the comb, first remove the scabs or crusts. Bathe the parts with a solution of one teaspoonful of creolin in one pint of water. When cleansed and dried apply ointment.

Crop Inflammation or Sour Crop.

79.—R.　Bismuth Subnitrate,　　　　15 grains.
　　　　Sodium Bicarbonate,　　　　　4 grains.
　　　　Water,　　　　　　　　　　　1 ounce.

Mix and give one teaspoonful after emptying the crop in cases of sour crop or crop inflammation not due to poisoning. Allow no food for one day.

Douglas Mixture.

80.—R.　Iron Sulphate,　　　　　　　16 ounces av.
　　　　Dilute Sulphuric Acid,　　　1 fluid ounce.
　　　　Water sufficient to make one gallon.

Mix and keep in stone jug or glass jar. One tablespoonful of this mixture may be given in each gallon of drinking water, allowing the birds no other drink. It will be found useful in all diseases where a cheap tonic is desired and is claimed to prove beneficial in cholera. Mix the sulphuric acid and water in an *open* vessel, add acid to the water.

Favus.

81.—R. Formaldehyde (40 per cent.), 10 drops.
        Sweet Oil, 4 ounces.

Mix.—For favus apply daily to moisten the scabs and crusts. When crusts are easily removed use a weaker solution made by adding more oil.

Feather Eating.

82.—R. Powdered Aloes, One level teaspoonful.
        Vaseline, Lard or other Fat, ½ pint.

Melt fat and stir in aloes, stirring until cool. Apply to feathers about the picked or bare area to prevent feather pulling.

Frost-Bite.

83.—R. Powdered Camphor, 45 grains.
        Wool Fat, 4 drachms.
        Hydrochloric Acid, 20 drops.
        Vaseline, 4 drachms.

Mix thoroughly. For frost-bite rub into frost-bitten parts when fowls are on roost.

Gape Remedy, No. 1.

84.—R. Iron Sulphate, 30 grains.
        Licorice Root, 2 ounces av.
        Red Pepper, 4 drachms av.
        Powdered Red Sandal-wood, 8 drachms av.
        Powdered Fenugreek, 8 drachms av.

Above ingredients should be powdered and thoroughly mixed, then rub into a stiff paste with a little New Orleans molasses. Make into pills or slugs about a quarter of an inch in diameter. Dissolve one of these pills in half a gallon of drinking water, allowing the chicks no other drink.

Gape Remedy, No. 2.

85.—R. Mix one tablespoonful of creolin in one quart of water. This mixture should be used in a spray pump throwing a very fine spray or mist. Confine the little birds in a box or barrel where they will be obliged to inhale the mist and spray the solution lightly over them, being careful not to use a sufficient amount to get them wet. It will set them coughing and sneezing with the result that the gape worms will be expelled. This remedy has been used in the southern states with considerable success. The solution also proves an effective remedy for catarrhal colds and contagious catarrh when sprayed around the affected birds while they are on the roost at night.

"Going Light" and Cholera Remedy.

86.—R.	Iron Sulphate,	1 ounce.
	Calcium Phosphate,	8 ounces.
	Fenugreek,	4 ounces.
	Black Pepper,	2 ounces.
	Capsicum (red pepper), ..	1 ounce.
	Sodium Chloride (table salt),	1 ounce.
	Calcium Carbonate (powdered chalk),	4 ounces.

Reduce all to fine powder and thoroughly mix. Give one teaspoonful in one quart of mash food.

Iodoform Ointment.

| 87.—R. | Finely-powdered Iodoform, | one part. |
| | Pure Vaseline, | 20 parts. |

Rub well together. This will be found useful in chicken pox, conjunctivitis, ulceration of the eyes or eyelids, ulceration of the vent, and for application to all running sores. First bathe the parts with a solution of one teaspoonful of creolin in a pint of warm water. Thoroughly cleanse and dry the parts, then apply the ointment.

Keratitis.

Ulceration of eyes.

| 88—R. | Yellow Mercury Oxide, | 2 grains. |
| | Vaseline, | 2 drachms. |

Rub together well. For ulceration of eyes in fowls apply a little between the lids daily.

Leg Weakness and Phlebitis.

| 89.—R. | Oil Cajaput, | 2 fluid drachms. |
| | Alcohol, | 8 ounces. |

Mix and rub legs with the mixture or apply on cotton in cases of leg weakness with hot swollen legs and enlarged veins.

Limberneck.

| 90.—R. | Oil of Turpentine, | 2 teaspoonfuls. |
| | Sweet Oil, | 1 tablespoonful. |

Mix and give all at one dose for limberneck. Follow in half an hour with a tablespoonful dose of ginger tea made by adding one teaspoonful of powdered ginger to a cupful of warm milk.

Liver Trouble—Diarrhoea.

91.—R. Calomel, ¼ ounce.
Sodium Bicarbonate, 1¼ ounces.
Sugar, 1 ounce.

Mix and rub together thoroughly. For liver trouble and yellow or green diarrhoea use for adult fowls one level teaspoonful in one quart of mash food, mixed with cold water. Do not use salt in mash when this remedy is used.

Piles.

92.—R. Powdered Fresh Raw Horse Chestnut
(Buckeye), one teaspoonful.
Powdered Fresh Witch Hazel Leaves,
one teaspoonful.
Vaseline, 1 ounce.

Rub together well to thoroughly mix. For protrusion of vent and piles of fowls apply daily and push back affected parts. If not effective, try iodoform ointment R. No. 87.

"Pip," No. 1.

93.—R. Potassium Chlorate, 20 grains.
Glycerine, ½ fluid ounce
Water, ½ fluid ounce.

Mix and apply freely to mouth as a wash in cases of mouth inflammation or "pip" with redness of mucous membrane and hard scale on tongue.

"Pip," No. 2.

94.—R. Boracic Acid, 15 grains.
Water, 1 ounce.

Mix and apply freely in cases of "pip" where a hard, dry shell forms on tongue. Do not try to remove the "pip" or dried membrane until it comes away readily.

Remedy for Worms.

95.—R. Oil of Turpentine,   two teaspoonfuls.
        Olive Oil or Sweet Oil,   one tablespoonful.

Mix and give the whole at one dose to fowls affected with worms after bird has been fasting from 6 to 12 hours. The remedy should be introduced directly into the crop by means of a rubber tube passed down the throat, and can be more easily given by aid of a common bulb or a piston syringe. In cases where large quantities of worms are found in the droppings, the dose may be doubled for adult fowls. For little chicks give from one-half a teaspoonful to two teaspoonfuls at one dose, according to age. Disinfect all droppings with strong creolin solution, R. No. 113.

This remedy is very effective for intestinal worms and has also proved useful in cases of gapes. Where chicks are affected with gapes, moisten a loop of horse hair with above mixture, shaking off any excess of oil, and introduce same into the windpipe of chick, holding the little bird's neck extended. Withdraw the loop of horse hair with a slightly twisting motion and you will find clinging to it the gape worms.

Roup Cure.

96.—R. Copper Sulphate (blue vitriol, coarse
          powdered),   4½ pounds.
        Potassium Nitrate (fine powdered),   ½ pound.
        Fuchsine (medicinal),   ½ ounce.
        Aniline Red, dye,   80 grains.

Mix thoroughly. This makes a good cheap roup cure similar to some of the commercial preparations and the cost of manufacture of above quantity should not be over $1.50 at present prices of drugs. All ingredients can be obtained through a wholesale druggist. CAUTION.—Blue vitriol is POISON.

DIRECTIONS.—This remedy is to be used in the drinking water. One-half a level teaspoonful to an ordinary 10-quart bucket of water is sufficient. A strong solution of the remedy may be made by using just enough water to dissolve the powder and this proves an effective application in canker and chicken pox. It is applied by means of a cotton swab (a piece of absorbent cotton twisted about the end of a sharp stick) barely moistened in the solution. Apply the moist swab to the injured parts and gently brush it over them.

Roup Dip.

97.—R. Luke Warm Water,   1 pint.
        Pure Creolin,   1 teaspoonful.

Mix well and use this remedy as a dip into which to dip the fowls' heads for birds affected with roup, roupy colds, canker, swollen head, inflammation of the eyes or conjunctivitis, and similar diseases. Do not dip more than eight or ten birds in the same pint of solution. Mix fresh as required.

Roup Lotion.

98.—R. Oil of Turpentine,   1 ounce.
        Glycerine sufficient to make   8 ounces.

Mix thoroughly. This remedy is useful for bathing face and eyes and injecting into the nostrils, also for swabbing out throat in roup and all roupy colds.

Roup Pills.

99.—R. Clear Lard,   1 tablespoonful.
        Vinegar,   2 teaspoonfuls.
        Cayenne Pepper,   2 tablespoonfuls. [tea]
        Mustard,   2 teaspoonfuls.

Mix above ingredients well together and add sufficient flour to make stiff dough. Roll into pills about three-eighths of an inch in diameter. Give one of these pills to each sick fowl at night. Repeat in 12 hours if necessary. The pills should be forced down the patient's throat.

Roup Powder.

100.—R. Licorice Root,   6 ounces.
        Potassium Chlorate,   4 ounces.
        Powdered Cubebs,   4 ounces.
        Powdered Anise,   2 ounces.

Reduce to fine powder and thoroughly mix. One heaping teaspoonful of this mixture may be given in 10 lbs. of mash food. Where fowls are too sick to eat it is seldom possible to effect a cure. This remedy has proved effective in a few such cases when made into pills with a little extract of gentian, just enough to work it up readily into pill form, making the pills about one-fourth of an inch in diameter. One of these pills may be given morning and night.

Sudden Colds, No. 1.

101.—R. Spirits of Camphor, 20 drops.
Sugar, 1 teaspoonful.

Mix and dissolve the whole in a pint of drinking water. Allow the bird no other drink. Useful for all sudden colds having snuffles, watery or frothy eyes, running at the nostrils and no odor.

Sudden Colds, No. 2, For Show Birds.

102.—R. Adrenalin Chloride (P. D. & Co.
1 to 1000 solution), 2½ fluid drachms.
Chloretone, 5 grains.
Distilled Water sufficient to make 1 fluid ounce.

Mix and use in atomizer throwing fine spray. This remedy is particularly useful for valuable show birds that take cold in the show room or during shipment. The mixture may be sprayed directly upon the mucous membrane of throat and into the nose twice daily, and will rapidly dry up the discharge. The preparation is expensive and is recommended for show birds or exceptionally valuable specimens only.

Show Bird Stimulant.

103.—R. Old Jamaica Rum, 4 fluid ounces.
New Orleans Molasses, 4 fluid ounces.

MIX.—For show birds sent by express in cold weather give one teaspoonful when shipped and another on arrival. Prevents taking cold and keeps bird quiet. Do not be in a hurry to feed bird immediately after cooping. Supply water and let him rest a while before feeding.

Tonic and Indigestion Remedy.

104.—R. Tincture of Red Cinchona, 1 fluid ounce.
Tincture of Chloride of Iron, 1 fluid drachm.
Tincture of Nux Vomica, 4 fluid drachms.
Glycerine and water equal parts
sufficient to make 4 fluid ounces.

MIX.—This remedy will be found useful as a general tonic and also in cases of indigestion. Give one teaspoonful in a quart of drinking water, allowing the birds no other drink.

Ulcers of Vent.

105.—R. Calomel, 10 grains.
         Lime Water, 1 fluid ounce.

Mix and shake well before using. For ulcers of vent apply freely on cotton. If not effective, try idoform ointment, R. No. 87.

Vent Gleet Ointment.

106.—R. Powdered Iodoform, ½ drachm.
         Lard, 1 ounce.

Melt the lard, add the iodoform, and stir until cool. The remedy is useful for applying daily in cases of vent gleet, ulceration or inflammation of vent; also in baldness and white comb.

CHAPTER VI.

TREATMENT OF POISONING.

Valuable fowls are frequently poisoned by having access to waste salt or old brine mixtures, commercial fertilizer, paint skins and spray mixtures used in the garden. The following formulæ will be found useful.

Poisoning with Arsenic.

Caused by eating Paris green, fly paper, London purple, etc.

107.—R.	Hydrated Oxide of Iron,	2 ounces.
	Calcined Magnesia,	4 drachms.

Mix. For arsenic poisoning, give above in teaspoonful doses, frequently repeated until relieved. Give chalk and water freely. Follow with white of egg, flaxseed tea or mucilage, or chalk and milk given freely. Give small doses of whiskey or brandy and milk. Give two teaspoonfuls of castor oil and keep fowl warm.

Poisoning with Copper.

From eating food contaminated with blue vitriol spray mixtures, or overdosing with commercial roup cures.

108.—R.	Common Hard Yellow Soap,	1 teaspoonful.
	Water to soften,	2 tablespoonfuls.

Mix. For copper poisoning, give above in teaspoonful doses and repeat if necessary. Feed white of eggs and milk, flaxseed tea or mucilage freely. Give stimulants.

Poisoning with Ergot.

From eating grain (usually rye), affected with spur or smut.

109.—R.	Tannin,	30 grains.
	Water,	1 fluid ounce.

Mix. For ergot poisoning, give above mixture freely or make fowl drink strong tea of white-oak bark. Give two teaspoonfuls of castor oil. Stimulate with milk and brandy or whiskey, and keep fowl warm.

Poisoning with Lead.

From eating paint skins, eating or drinking from old paint cans.

110.—R. Sodium Sulphate, 1 ounce.
Magnesium Sulphate, 2 ounces.
Water, 1 pint.

Mix. For lead poisoning, give in large doses, frequently repeated until relieved. Give large quantities of white of egg and flaxseed tea with milk.

Poisoning with Phosphorus.

From eating rat or roach poison.

111.—R. Copper Sulphate (powdered), 5 grains.
Water, 1 fluid ounce.

Mix. For phosphorus poisoning, give above remedy in two doses. Repeat if necessary. Give pills of powdered charcoal freely. One teaspoonful of spirits of turpentine mixed with mucilage may be given. Do not give oils or fat.

Poisoning with Salt, Strong Lye, Quick Lime or Soda Nitrate.

Caused by eating refuse from old brine, moist wood ashes, or strong lye, unslaked lime and fertilizer mixtures.

112.—R. Flaxseed Meal, 1 ounce.
Boiling Water sufficient to make thick mucilage.

Mix. For salt, lime, lye or soda poisoning, give above remedy freely, together with milk and stimulants like whiskey or brandy. Strong black coffee is often useful. Weak vinegar and water may be given.

CHAPTER VII.

DISINFECTANTS.

It is always a good plan for the poultryman to have on hand a supply of a reliable disinfectant for use on his plant. Buildings, incubators and brooders all require occasional disinfecting to keep them sweet and wholesome. Undoubtedly the best all around disinfectant for poultrymen is pure creolin, and I prefer to use Creolin-Pearson, although same is more expensive than the many substitutes offered for sale for veterinary purposes. Among the most reliable creolin substitutes are zenoleum, Buffalo sanitary fluid, napcreol, sulpho-nathol, and Liquor Cresolis Compositus, U. S.

Creolin Solution (Strong).

113.—R. Creolin-Pearson, 6½ fluid ounces.
Water, 1 gallon.

Mix. This solution makes a strong disinfectant for all farm purposes, and it is more potent in germ destruction than a 1 to 1000 solution of corrosive sublimate. It is practically non-poisonous. It may be used to disinfect incubators or as a spray, paint or wash to disinfect woodwork and floors in poultry buildings, or for any other germ killing, cleansing and deodorizing purposes in sink drains, vaults, outbuildings, etc.

Creolin Solution (Medium).

114.—R. Creolin-Pearson, 3 fluid ounces.
Water, 1 gallon.

This is a weaker creolin solution than R., No. 113, and is very effective where a less powerful disinfectant is needed. Recommended for disinfecting incubators and brooders.

Colored Whitewash.

115.—R. Unslaked lime, one-half bushel. Slake in a barrel with sufficient hot water to cover lime four or five inches. Stir briskly until thoroughly slaked. When completely slaked, add 2 lbs. sulphate of zinc and 2 lbs. of table salt dissolved in hot water. It may be colored with Venetian red, India red, ultramarine blue, umber, ivory-black or yellow ochre, to any tint desired. Wash should be applied hot with a good brush.

Formaldehyde Disinfectant.

116.—R. Formaldehyde (40 per cent.), 1 pint.
Water, 2 gallons.

This mixture may be used for poultry house disinfection. The fowls should be shut out of the building, which should have its interior thoroughly sprayed and all windows and doors closed, and allowed to remain closed to confine the vapors for a few hours. Air out well before birds are allowed to use the building.

Kerosene Emulsion.

117.—R. Yellow Soap, 1 pound.
Boiling Water, 1 gallon.
Kerosene, 2 gallons.

Shave soap fine and dissolve in boiling water. When soap is dissolved, add kerosene while mixture is still hot. Stir or churn briskly until creamy emulsion is formed, then add sufficient cold water to make fifteen gallons. It may be applied to poultry buildings with a spray pump, brush or sprinkling pot to drive away lice and mites.

Sulphuric Acid Disinfectant.

118.—R. Sulphuric Acid (50 per cent. sol.), 16 ounces.
Water, 6 gallons.

Have the water in a wooden tub or barrel. Add the dilute sulphuric acid very slowly to the water. Be careful that it does not splash on to hands or face.

Always use diluted sulphuric acid, not more than 50 per cent. strength for making above solution. If sulphuric acid gets on flesh or clothing it will burn quickly and deeply. In accidental burning, large quantities of water should be applied at once, and as soon as possible a solution of baking soda or ammonia should be applied to neutralize the acid. Carefully and cautiously used, sulphuric acid makes a valuable and cheap disinfectant. When diluted, as advised in prescription, it may be safely used for sprinkling runs, saturating floors and walls of poultry buildings. Also for disinfecting droppings. The sulphuric acid mixture should not be prepared in glass bottles, stone jugs, or iron vessels. After mixing it may be kept in jugs or bottles if desired.

Whitewash for Spraying, No. 1.

119.—R. Slake quick-lime in sufficient warm water to make a thin creamy mixture. To each gallon of this wash add 2 fluid ounces of *crude* carbolic acid. Strain through burlap or fine sieve into spray pump.

Whitewash for Spraying, No. 2.

120.—R. A good whitewash for spraying may be made by slaking quick-lime in just sufficient water to make a thick paste. Add a pint of melted lard or other grease and a cupful of common table salt to each half bushel of lime while slaking. This lime paste may be stored in tubs or barrels and will keep perfectly as long as kept moist and covered. Before using, add to lime paste a sufficient amount of water to give the wash the consistency of thin cream. Strain through burlap or a fine sieve into a spray pump. It may also be applied with an old broom or coarse brush, slapping it on freely so that it will work well into cracks.

Whitewash That Will Not Rub.

121.—R. Slake one peck of quick-lime in boiling water, keeping just covered by water while slaking. Strain through coarse cloth, add 2 quarts of fine salt dissolved in warm water, one pound of rice meal boiled in water to a thin paste; one-quarter of a pound of whiting; one-half pound of glue dissolved in warm water. Mix all thoroughly and let stand covered for two or three days. Stir occasionally. Heat the mixture before using.

CHAPTER VIII.

EGG PRESERVATIVES.

There are many preparations recommended for preserving eggs. Of these, undoubtedly the sodium silicate or water-glass solution is the most satisfactory. All eggs intended to be saved should be clean and strictly fresh when packed. Preferably use eggs from hens where there is no male bird in the flock.

Water Glass Method.

122.—R. Sodium Silicate (water-glass syrup), 1 quart.
Boiled Water, 9 quarts.

Mix. When solution is cool, it is ready to use. Eggs should be packed in stone or glass jars, galvanized iron tanks, clean wooden kegs or other clean containers that are easy to handle and can be kept well covered. Pack eggs in layers, small end down. Pour over them dilute water-glass solution until it stands two or three inches above the topmost layer of eggs. Cover container and set in cold place until eggs are wanted for use. This solution will keep eggs in quite good condtion for from six to ten months. Preserved eggs intended for boiling should have the large end pierced with a darning needle before cooking to prevent bursting of egg. Water-glass syrup may be had of your druggist at a cost of about 65 cents to $1.00 per gallon.

Lime Water Brine Method.

123.—R. Quick-Lime, 16 ounces.
Boiled Water, 1 gallon.
Table Salt, 8 ounces.

Slake lime thoroughly in water, then add the salt. Stir mixture well and let stand for a day or two. Draw off the clear solution. Eggs should be packed in clean receptacles small end down and covered with lime water brine solution until it stands three inches or more above topmost layer of eggs. Cover tightly and place in a cool cellar until eggs are needed for use.

Southern Lime Method.

124.—R. Unslaked Lime, 6 pounds.
Salt, 2 pounds.
Water to thoroughly slake above. 12 gallons.

Stir until well slaked. Let stand until perfectly clear. Draw off 10 gallons of clear salt and lime water. Now add the following:

Baking Soda, 2½ ounces.
Cream of Tartar, 2½ ounces.
Saltpetre, ¼ ounce.
Borax, ¼ ounce.
Alum, ½ ounce.

Mix and reduce above ingredients to a fine powder, dissolve in 2 quarts of boiling water and add to the 10 gallons of salt and lime water. The eggs should be packed in a clean, watertight barrel or half barrel. Cover the eggs with the preserving fluid and let water stand one or two inches above topmost layer of eggs. Put a clean muslin cloth over top of eggs inside of barrel. Do not let edges of cloth hang outside. On top of cloth, place the moist paste or settlings left from slaking the lime. The eggs must be kept covered with the preserving fluid. It is claimed that this method will keep eggs in good condition for six to ten months. Eggs preserved in this manner are apt to have a sharp alkaline taste, and are not as desirable as those preserved by the water-glass method.

CHAPTER IX.

MISCELLANEOUS.

In this chapter are given useful formulae which could not be properly classified in the preceding chapters but which will be found of practical use on the poultry plant.

Cement for Floors.

125.—R. Best Portland Cement,   1 bushel.
         Coarse Sharp Sand,   2 bushels.
         Water sufficient to make a thick, stiff mixture.

The cement must not be wet or sloppy and should be well tamped or beaten down with a flat spade or plank. This cement should be built up on a foundation of crushed rock or small stones. Top off with a finishing coat about one inch thick of equal parts by measure cement and coarse sand.

Clover Hay.

126.—Mow clover after the sun is up and the dew thoroughly dried off. Plan your mowing for a good, warm, sunny, breezy day. Let lie until afternoon, then make into small loose cocks and cover with canvas to protect from dew. Spread in morning to let dry in hot sun and breeze. When leaves are nearly dry bunch up into small cocks before dew falls. Hay should lay as loosely as possible in the cocks. On the next day open out the cocks into small bunches. To avoid loss of leaves do not shake out. Be sure that hay is thoroughly cured before placing in barn.

Clover Rowen.

127.—Second-crop clover or rowen is best for poultry and should be cut when about three to four inches high. Mow early in the morning as soon as dew is dry. Let lay until evening, turning over three or four times during day. Toward night gather up into a small stack and cover with canvas to keep off dew or place under shed or in barn. It should be dried thoroughly by spreading in the sun. Care should be taken not to allow it to get wet with dew or rain. Continue drying daily and stacking at night until thoroughly cured. Put in a loose stack for 24 hours. If it begins to sweat turn it over and thoroughly dry. Stack again and repeat the process if necessary. The clover when thoroughly cured is ready to cut into short lengths, one-quarter to one-half inch long in a clover cutter. It may be stored in bins or in sacks. If thoroughly cured it will keep for a long time without becoming musty or moldy. Do not shake or handle clover too frequently while curing, as you wish to retain the leaves.

Hen Manure Fertilizer.

128.—R.
Dry Hen Manure,	30 lbs.
Sawdust.	10 lbs.
Acid Phosphate (fertilizer),	16 lbs.
Kainit,	8 lbs.

Mix and thoroughly pulverize. It should be used sparingly for all garden crops. Particularly useful for strawberry beds, asparagus, corn and tomatoes.

Hen Manure for General Fertilizer.

129.—R.
Poultry Manure,	200 lbs.
Pulverized Gypsum or Land Plaster,	30 lbs.
Fine Loam,	600 lbs.

Mix thoroughly, moistening same a little from time to time and working over frequently until you have a well pulverized mass. This may be spread broadcast on grass land or used on the garden as you would other manure. A very satisfactory mixture for fertilizing lawns and for home garden crops. Poultry manure may also be used by mixing with horse manure, cow dung and sawdust and allowing pigs to work same over, using the same as ordinary stable manure.

Poultry Manure and Wood Ashes.

130. Mix together equal parts by bulk poultry manure, wood ashes and fine peat or loam. Throw the mixture into little heaps under cover and moisten occasionally by sprinkling with water. Turn over now and then until thoroughly mixed, fine and light. It may be used in the same manner as commercial garden fertilizers for all common crops.

STORING VEGETABLES FOR WINTER FEEDING.

New Jersey Method.

131. In localities where the soil is sandy and winters are mild, make a long trench about one foot deep on a dry, sandy ridge; line with straw. Fill in trench with cabbages, beets, mangels and other vegetables intended for winter use. Cover lightly with straw or hay and mound up with four or five inches of sand. In using vegetables begin to use from one end of the trench.

Northern Method.

132. Dig a deep pit in a gravelly or sandy hillside where land is well drained. Line pit with bedding hay or straw. Fill in with vegetables, cover with straw, placing a few plank at front of pit to make same easily accessible. Cover with leaves, hay or straw, and bank up with earth 10 to 18 inches deep on top and sides.

Storage in Cellars.

133. In cellars where vegetables do not freeze place potatoes loose in bins or store in burlap sacks. Beets, turnips, mangels and carrots should be placed in bins and covered with dry sand or earth. Cabbages may be hung up by the roots or stacked in piles in one corner of the cellar.

Homoeopathic Poultry Remedies

Introduced by me to poultrymen in 1893 and their value proved in daily successful use by experienced poultrymen. Used in the drinking water. Will not lose strength or deteriorate with age.

REMEDY No. 1.—For indigestion, sour crop, liver disease, constipation, limberneck.

REMEDY No. 2.—For chicken-pox, chronic catarrh and all diseases where pus is present.

REMEDY No. 3.—For catarrhal colds, bronchitis, croup, rattling in throat.

REMEDY No. 4.—For roup, roupy colds, snuffles, discharge from nostrils and eyes.

REMEDY No. 5.—For diarrhoea and cholera.

REMEDY No. 6.—For lameness, swollen legs, rheumatism and cramps.

REMEDY No. 7.—For intestinal parasites and worms.

REMEDY No. 8.—For diseases of egg organs, infertility, soft-shelled eggs.

REMEDY No. 9.—Conjunctivitis, swollen eyes, blindness in chicks.

REMEDY No. 10.—Useful in diphtheria, canker and vent gleet.

Put up in neat bottles with directions on label. Price, 35 cents each single remedy, postpaid, or any three remedies for $1.00. All ten remedies to one address by express, purchaser to pay express charges, $2.50. ORDER BY NUMBER.

White Diarrhoea

My method of Successful Chick Rearing and "Facts About White Diarrhoea" with special prescription "D135" for the prevention and cure of so-called White Diarrhoea will help stop your losses and save your chicks. The book and the remedy both to one address will be sent postpaid on receipt of one dollar. White Diarrhoea remedy alone, without book, 50 cents, postpaid.

Address all orders to

Dr. P. T. Woods, ~~Box 126, Middleton, Mass.~~

32 Woodward Ave.,
Buffalo, N.Y.

CPSIA information can be obtained
at www.ICGtesting.com
Printed in the USA
BVHW031107200622
640190BV00008B/113